## I. Introduction

The usefulness of scanner data for analyzing the retail sector is widely seen as a "success story" by both academics and industry participants (Bucklin and Gupta 1999). It remains an open question, however, whether aggregate-level data can reliably be used to estimate the demand for a set of products, or if store-level data is required. Although previous research shows that demand estimates based on aggregate data are biased when stores are heterogeneous, only partial solutions to the aggregation bias problem have been developed thus far.

We propose a methodology for avoiding aggregation bias that allows inter-store heterogeneity to be explicitly controlled for with aggregate data. This is accomplished by exploiting information regarding the distribution of store characteristics, information that is only partially utilized in extant aggregate demand models. Our approach is highly practical since it relies solely on standard scanner data of the type produced by the major vendors, ACNielsen and Information Resources, Inc. (IRI).

Throughout this paper, "aggregate-level" refers to data where the sales from multiple stores are combined. Examples of aggregate datasets include city-level data (e.g., all supermarkets in Chicago), and city-chain data (e.g., all of Jewel's supermarkets in Chicago).[1] Researchers who lack access to store-level data must depend on these types of datasets to estimate the demand for a set of products. Unfortunately, aggregate-level scanner datasets do not report the marketing-mix characteristics of products at each store, such as their price and promotional activity. Unable to model inter-store heterogeneity with such data, researchers have employed a representative store paradigm where each consumer faces the average price and promotion level across all stores. Recent examples include Nevo (2000a), Hausman and Leonard (2002), Cotterill and Samson (2002), and Perloff and Ward (2003). Modeling aggregate

---

[1] A confidentiality agreement with ACNielsen prohibits retailer names from being revealed. This example does not indicate whether the dataset employed contains the Jewel supermarket chain in Chicago.

demand in this simplified manner comes at a high cost, however. The aggregation bias literature demonstrates that when stores have heterogeneous marketing-mix strategies, the representative store model produces biased demand estimates (Christen et al. 1997).

Researchers continue to rely on the representative store paradigm since they lack a better alternative. The two solutions to the aggregation bias problem that have been developed hereto each have significant limitations. The method proposed by Christen et al. (1997) has informational requirements that go beyond what is reported in standard scanner datasets, preventing it from being widely applied. Link (1995) suggests an alternative solution for avoiding aggregation bias that does not require special data. Link demonstrates that disaggregating the data by each product's promotional activity reduces aggregation bias. Aggregation bias is not completely avoided, however, as stores with the same promotional activity for a given product may have different promotions for the other items they carry.

We overcome this shortcoming by extending the univariate (own-product) promotional decomposition advocated by Link to a multivariate decomposition that accounts for both own- and cross-product promotional activity. This is accomplished by exploiting promotional distribution information that is typically reported in aggregate-level scanner datasets. Specifically, we more fully utilize data on the fraction of stores where a product has a given level of promotional activity (e.g., an in-store display or an advertisement in a weekly circular). Further, aggregate scanner datasets separately report each product's price for each type of promotional activity. We show how to use this information to go beyond the representative store model.

Heterogeneous stores are allowed for, where each store type is a unique combination of promotional activity for each product. We obtain an aggregate demand model that is consistent with store-level heterogeneity by adding up demand for each product across each store type. This formulation requires knowledge of the joint distribution of promotions across stores so that we can calculate the number of stores of each type. We estimate this joint distribution using the univariate distribution of each product's promotional activity, information that is reported in the scanner datasets produced by the two major vendors.

We demonstrate the advantages of our proposed methodology by estimating the demand for super-premium ice cream. Using a random coefficients logit demand model, we not only show that our framework produces sensible results, but estimates that are measured more precisely than in the standard, representative store model. In contrast, the traditional model produces implausible estimates due to aggregation bias. These findings are corroborated by Monte Carlo analysis that shows our promotional disaggregation approach substantially outperforms the representative store framework.

The paper is organized as follows. Section two reviews the methods that have previously been developed to avoid aggregation bias. Section three presents a consumer demand model that accounts for marketing-mix heterogeneity across stores, and then uses this framework to estimate the demand for super-premium ice cream. In section four, we employ Monte Carlo analysis to compare our disaggregated promotion framework to the standard representative store model. Section five concludes.

## II.  Aggregation Bias

Estimating demand with aggregate data often leads to model mis-specification, or "aggregation bias." This section reviews two previously developed solutions for avoiding aggregation bias. Since this problem is widely known, we do not detail why aggregation bias occurs. For a detailed consideration of this issue, Theil (1954) and Krishnamurthi et al. (1990) analyze the linear model; Lewbel (1992), Christen et al. (1997), and Chung and Kaiser (2000) analyze the constant elasticity model; and Allenby and Rossi (1991) and Krishnamurthi et al. (2000) analyze the logit model.

Link (1995) argues that data aggregation across stores with heterogeneous marketing activity is the most significant source of bias in practical applications. Link suggests that aggregation bias be avoided by employing data that has been aggregated across stores with homogeneous marketing activity.[2] However, even if one obtains data that is aggregated across stores where a product's own promotions are homogenous, heterogeneity in the promotions of competing products may remain. Thus, Link's approach does not account for aggregation bias in cross-product effects. A further limitation is that it requires demand for each product to be separately estimated, since it is possible to disaggregate by promotional activity for only one product at a time. This prevents it from being applied to certain frameworks, such as the random coefficients logit model where the demand for each product is jointly estimated (Nevo 2000b).

Christen et al. (1997) propose a methodology to "de-bias" demand estimates based on aggregate data. First, demand is estimated using simulated store-level data that has been aggregated across stores. The average difference between the true and estimated parameters from the simulation is then added to the estimates from an empirical application, to de-bias the results. It can be difficult to estimate the magnitude of aggregation bias reliably as one may have insufficient information to calibrate the simulated data to the actual data. The de-biasing procedure may not eliminate aggregation bias and if done poorly could exacerbate the problem. Moreover, due to having data requirements that go beyond what is reported in standard datasets, this procedure has not been widely employed.

## III. Demand Model

While researchers are aware of the problem posed by aggregation bias, thus far they have developed only limited solutions. Motivated by this gap in the literature, we develop a consumer demand model that accounts for marketing-mix heterogeneity across stores. We then add up the demand for each product across consumers to obtain an internally consistent aggregate demand model.

---

[2] Boatwright et al. (2004) is a recent example that uses this methodology to avoid aggregation bias.

Data

We utilize supermarket scanner data provided by ACNielsen for the super-premium ice cream category. The dataset separately reports weekly sales for 14 city-chain combinations for the period December 1998 to June 2001 (132 weeks). However, to avoid complications involving entry into certain geographic areas, only a subset of the data is used; we analyze the last 80 weeks of data for the 11 city-chain combinations where the same four brands comprise the entire category. To comply with a confidentiality agreement with ACNielsen, they are referred to as Brand A, B, C, and D.

The data separately reports unit and dollar sales for four mutually exclusive levels of promotional activity $m \in M$, where $M$ = {"No Promotion," "Display Only," "Feature Only," "Feature & Display"}. A "Feature" is a print advertisement, such as in a promotional circular, while a "Display" is a secondary sales location within a store used to draw special attention to a given product. The demand specification presented below details the conditions under which consumer demand can be added up across the subset of stores where a product has a given type of promotional activity, so that it can be consistently estimated with aggregate data. The required conditions are less restrictive than those needed to perform an equivalent aggregation exercise using a representative store model.

In the super-premium ice cream category each brand's UPCs represent a different flavor, with a particular flavor rarely available for more than one brand (e.g., "Chunky Monkey" is available only for the Ben & Jerry's brand).[3] The large number of idiosyncratic flavors limits the usefulness of this characteristic for estimating substitution patterns. Other meaningful characteristics are common across UPCs for a given brand; each brand's UPCs share the same brand image and, within any given store, they are identically priced and promoted. Below, we

---

[3] In addition, UPCs vary by package size. Most brands of super-premium ice cream are available only in pint-sized containers, however, with larger package sizes representing a small fraction of category sales. We therefore omit them from the analysis by restricting the dataset to pint-sized cartons.

develop a product-level demand specification. Nonetheless, since the control variables employed vary only by brand, this specification simplifies to a brand-level demand model.

## Disaggregated Promotion Model

The following details the random coefficients logit demand model employed in the empirical analysis. In every time period $t$, each consumer $i$ purchases that item which generates the highest utility. The choices are the set of currently available products $J_t$ or the "outside good." We normalize the utility derived from purchasing the outside good to a mean utility of zero, $U_{i0t} = \omega_{i0t}$, where $\omega_{i0t}$ is i.i.d. Type I Extreme Value. For the remaining choices, consumer $i$'s utility for product $j$ during week $t$ is determined by its promotional activity $m_{ijt} \in M$, price $p_{ijt}$, a set of product characteristics $X_{ijt}$ that has an associated vector of random coefficients $v_i$, a set of additional controls $Z_{jt}$, and an i.i.d. error term $\omega_{ijt}$ that is distributed Type I Extreme Value.[4]

$$(3.1) \quad U_{ijt} = \mu^{m_{ijt}} + \beta^{m_{ijt}} p_{ijt} + X_{ijt} v_i + Z_{jt} \gamma + \omega_{ijt}$$

Product characteristics $X_{ijt}$ include a set of dummy variables for each brand, price $p_{ijt}$, and dummy variables for "Display Only," "Feature Only" and "Feature & Display." Control variables $Z_{jt}$ consist of brand fixed effects for each city-chain combination, a fourth order time trend, the number of products available in the category and the square of this variable (Ackerberg and Rysman, Forthcoming).

We estimate a separate intercept $\mu^m$ and price coefficient $\beta^m$ for each type of promotional activity $m \in M$. In light of conflicting empirical evidence regarding whether promotional activity makes demand more or less elastic (Blattberg et al. 1995), this specification allows for either possibility.

---

[4] City-chain subscripts are dropped from the control variables for ease of notation.

The model accommodates heterogeneity in consumer preferences through random coefficients $v_i$. We assume $v_i$ is mean-zero and i.i.d. Normally distributed with a block diagonal variance matrix $V = \begin{bmatrix} V_1 & 0 \\ 0 & V_2 \end{bmatrix}$. Denote the probability distribution function of $v_i$ by $\phi(v_i; V)$. $V_1$ corresponds to the brand dummy variables contained in $X_{ijt}$ (the fixed characteristics), while $V_2$ corresponds to the remaining price and promotion variables (the variable characteristics). We place no restrictions on $V_1$ and $V_2$ apart from the requirement that each be a symmetric positive semi-definite matrix. To keep the number of estimation parameters manageable, however, we assume the two sets of random coefficients are independently distributed.

While aggregate scanner datasets separately report each product's average price for each type of promotional activity $m$, they do not report any information regarding the distribution of prices across stores with the same promotional activity. Since the data lacks the information necessary to model this price heterogeneity, we assume that stores with the same promotional activity for a given product all charge the same price.

(3.2) $\quad p_{ijt} = p_{jt}^m, \forall i : m_{ijt} = m$

Condition (3.2) is substantially weaker than the price assumption usually made when estimating aggregate demand, where prices are assumed identical across all stores regardless of the level of promotional activity (e.g., Nevo 2000a). We recognize that even this weaker requirement may still be restrictive. In section four, we analyze whether its violation leads to aggregation bias.

Rather than relying on a representative store framework, we allow heterogeneous promotions across stores. Store type $g \in G$ is a vector containing each product's promotional activity. Consumer $i$ in week $t$ visits store type $g_{it} = \{m_{ijt}\}_{j \in J_t}$. Denote the element of $g_{it}$ that corresponds to product $j$ by $m_j(g_{it})$. Product characteristics $X_{ijt}$ are written as $X_{jt}(g_{it})$ since the included variables vary only by store type $g$. This allows us to re-write the utility function as follows.

$$(3.3) \quad U_{ijt} = \mu^{mj(g_{it})} + \beta^{mj(g_{it})} p_{jt}^{mj(g_{it})} + X_{jt}(g_{it})v_i + Z_{jt}\gamma + \omega_{ijt}$$

Apart from heterogeneity in price and promotional activity, all stores are identical. In addition, we assume each consumer is randomly matched to a store. This allows us to integrate over the distribution of random coefficients $v_i$ for the subset of consumers who visit a given store type $g$, even though aggregate scanner datasets contain no information regarding individual stores or consumers.

Each consumer $i$ purchases product $j$ only when that item generates the highest utility from among the available choices. The distributional assumptions provided above imply the following for $\hat{q}_{jt}^g$, predicted unit sales for product $j$ across all stores that have promotional activity $g$ during week $t$.

$$(3.4) \quad \hat{q}_{jt}^g = Q\pi_{gt} \int \frac{e^{[\mu^{mj(g)} + \beta^{mj(g)} p_{jt}^{mj(g)} + X_{jt}(g)v_i + Z_{jt}\gamma]}}{1 + \sum_{k \in J_t} e^{[\mu^{mk(g)} + \beta^{mk(g)} p_{kt}^{mk(g)} + X_{kt}(g)v_i + Z_{kt}\gamma]}} \phi(v_i; V)dv_i$$

$Q$ represents the total number of consumers in the market, while $\pi_{gt}$ is the fraction of consumers who visit store type $g$ in week $t$. Researchers often choose a value for $Q$ based on each market's population. We cannot do so here because our data is aggregated to the city-chain level, and we have no information regarding the number of people who frequent each chain. Therefore, we instead assume that $Q$ equals 10 times each city-chain's average category sales.

As we mentioned earlier regarding super-premium ice cream, product-specific characteristics cannot be meaningfully incorporated into the demand specification. Apart from idiosyncratic flavor differences, within any given store product characteristics such as price and promotional activity are identical across all UPCs for the same brand. For this reason we aggregate (3.4) to the brand-level, where $B$ denotes the set of brands and $\hat{q}_{bt}^g$ is predicted unit sales for brand $b$ across stores with promotional activity $g$ in week $t$.

$$(3.5) \quad \hat{q}_{bt}^g = \sum_{j \in J_t: b_j = b} \hat{q}_{jt}^g = Q\pi_{gt} \int \frac{\sum_{j \in J_t: b_j = b} e^{[\mu^{mj(g)} + \beta^{mj(g)} p_{jt}^{mj(g)} + X_{jt}(g)v_i + Z_{jt}\gamma]}}{1 + \sum_{\tilde{b} \in B} \sum_{j \in J_t: b_j = \tilde{b}} e^{[\mu^{mj(g)} + \beta^{mj(g)} p_{jt}^{mj(g)} + X_{jt}(g)v_i + Z_{jt}\gamma]}} \phi(v_i; V) dv_i$$

All variables in (3.5) that have a product $j$ subscript are identical across products with the same brand $b$. To simplify notation we therefore replace each $j$ subscript with a $b_j$ subscript. Equation (3.5) reduces to the following, where $N_{bt}$ denotes the number of products available in time $t$ that are part of brand $b$'s product line.

$$(3.6) \quad \hat{q}_{bt}^g = Q\pi_{gt} \int \frac{N_{bt} e^{[\mu^{mb(g)} + \beta^{mb(g)} p_{bt}^{mb(g)} + X_{bt}(g)v_i + Z_{bt}\gamma]}}{1 + \sum_{\tilde{b} \in B} N_{\tilde{b}t} e^{[\mu^{m\tilde{b}(g)} + \beta^{m\tilde{b}(g)} p_{\tilde{b}t}^{m\tilde{b}(g)} + X_{\tilde{b}t}(g)v_i + Z_{\tilde{b}t}\gamma]}} \phi(v_i; V) dv_i$$

To account for our transformation from a product- to a brand-level demand model, we update the definition of store type $g$. We now define $g$ as the set of promotional activity across the four brands. Since there are four brands and four types of promotional activity, $G$ contains $4^4 = 256$ unique store types. This simplification is possible since, as we discussed earlier, in the super-premium ice cream category each brand's entire product line is identically promoted within any given store.

Calculation of $\hat{q}_{bt}^g$ in equation (3.6) requires integration over the random coefficients $v_i$. One way to do so is to generate a random sequence of draws $\{v^l\}_{l=1}^L$ that are mean-zero and i.i.d. Normally distributed with variance matrix $V$. We then approximate equation (3.6) as follows.

$$(3.7) \quad \hat{q}_{bt}^g \approx \frac{Q\pi_{gt}}{L} \sum_{l=1}^{L} \frac{N_{bt} e^{[\mu^{mb(g)} + \beta^{mb(g)} p_{bt}^{mb(g)} + X_{bt}(g)v^l + Z_{bt}\gamma]}}{1 + \sum_{\tilde{b} \in B} N_{\tilde{b}t} e^{[\mu^{m\tilde{b}(g)} + \beta^{m\tilde{b}(g)} p_{\tilde{b}t}^{m\tilde{b}(g)} + X_{\tilde{b}t}(g)v^l + Z_{\tilde{b}t}\gamma]}}$$

Generating $\{v^l\}_{l=1}^{L}$ using a Halton sequence is a more efficient means of calculating $\hat{q}_{bt}^{g}$ (Train 1999). Since the Halton sequence produces values that are more smoothly distributed over the support of the Normal distribution than would occur under random sampling, we can choose a much smaller value for $L$ and still obtain accurate results. Nonetheless, we set $L = 1,000$ to make equation (3.7) as precise a representation of equation (3.6) as reasonably possible.[5]

Since our objective is to add up demand to the same level of aggregation as the available data, we calculate predicted unit sales $\hat{q}_{bt}^{m}$ by summing across all stores where brand $b$ has promotional activity $m$.

$$(3.8) \quad \hat{q}_{bt}^{m} = \sum_{g \in G: m_b(g)=m} \hat{q}_{bt}^{g}$$

This completes our demand specification. We have shown how to aggregate a consumer-level demand model with marketing-mix heterogeneity across stores.

Representative Store Model

To provide a point of comparison to the disaggregated promotion model described above, we also consider a counterpart model based on the representative store paradigm. Utility function (3.1) is replaced by the expected utility when each store charges $p_{jt}$, the average price across all types of promotions $m$. Previously, we included random coefficients for the price and promotional activity at the store consumer $i$ frequents. In the representative store model these variables are replaced by average price $p_{jt}$ and the percentage of stores with each type of promotion $m$, which we denote by $\pi_{jt}^{m}$.[6] As before, we include an i.i.d. error term $\omega_{ijt}$ that is distributed Type I Extreme Value. These changes lead to the following utility function for the representative store model.

---

[5] Following Train (1999), we discard the first 10 elements of the Halton sequence since early elements tend to be correlated. Therefore, we generate a sequence of 1,010 elements, rather than 1,000.

[6] In the representative store model we drop the $i$ subscript from $X_{ijt}$ since the included variables are identical across all consumers.

(3.9) $\tilde{U}_{ijt} = E(U_{ijt} \mid p_{ijt} = p_{jt}, v_i, X_{jt}, Z_{jt}) + \omega_{ijt} = \bar{\mu}_{jt} + \bar{\beta}_{jt} p_{jt} + X_{jt} v_i + Z_{jt} \gamma + \omega_{ijt},$

where $\bar{\mu}_{jt} = \sum_{m \in M} \mu^m \pi_{jt}^m$ and $\bar{\beta}_{jt} = \sum_{m \in M} \beta^m \pi_{jt}^m$

Utility function (3.9) implies the following characterization of predicted unit sales for product $j$ in week $t$.

(3.10) $\hat{q}_{jt} = Q \int \frac{e^{[\bar{\mu}_{jt} + \bar{\beta}_{jt} p_{jt} + X_{jt} v_i + Z_{jt} \gamma]}}{1 + \sum_{k \in J_t} e^{[\bar{\mu}_{kt} + \bar{\beta}_{kt} p_{kt} + X_{kt} v_i + Z_{kt} \gamma]}} \phi(v_i; V) dv_i$

Since the control variables employed vary only by brand, we replace each $j$ and $k$ subscript in (3.10) with a $b_j$ and $b_k$ subscript and then aggregate to the brand-level. As before, $N_{bt}$ denotes the number of items available in time $t$ that are part of brand $b$'s product line.

(3.11) $\hat{q}_{bt} = \sum_{j \in J_t : b_j = b} \hat{q}_{jt} = Q \int \frac{N_{bt} e^{[\bar{\mu}_{bt} + \bar{\beta}_{bt} p_{bt} + X_{bt} v_i + Z_{bt} \gamma]}}{1 + \sum_{\tilde{b} \in B} N_{\tilde{b}t} e^{[\bar{\mu}_{\tilde{b}t} + \bar{\beta}_{\tilde{b}t} p_{\tilde{b}t} + X_{\tilde{b}t} v_i + Z_{\tilde{b}t} \gamma]}} \phi(v_i; V) dv_i$

In the representative store model $\hat{q}_{bt}$ is defined as predicted unit sales across *all* stores. This contrasts with the level of aggregation employed in the disaggregated promotion model. The counterpart variable in that model is $\hat{q}_{bt}^m$, predicted unit sales across the subset of stores where brand $b$ has promotional activity $m$. Despite this difference, the representative store framework is a special case of the disaggregated promotion model. The two models are identical when all consumers observe the same price and promotional activity. Our disaggregated promotion model therefore does not offer any advantage when this condition holds, such as when demand is estimated with store-level data.

Differences arise, however, when demand is estimated with data aggregated across heterogeneous stores. Consumers visit stores with heterogeneous price and promotional activity in the disaggregated promotion model, while all stores are assumed identical in the representative store model. The disaggregated promotion model requires that stores with the same promotional activity for a given product all charge the same price. The representative store model more stringently assumes that every consumer observes the same price. The disaggregated promotion model recognizes that if a product is on promotion in 20% of stores, then 20% of consumers observe the promotion and 80% do not. In contrast, the representative store model assumes that consumers observe the average promotional activity across all stores (i.e., everyone observes a "partial" promotion). To summarize, the representative store framework ignores inter-store heterogeneity, averaging over differences in price and promotional activity. This approach is problematic since previous research demonstrates it leads to aggregation bias. In contrast, our disaggregated promotion framework explicitly models heterogeneous store types.

Data Requirements

Estimation of the disaggregated promotion model requires only standard scanner data of the type produced by the major vendors, ACNielsen and IRI. Typically, such data separately reports dollar and unit sales for four (mutually exclusive) types of promotional activity: "No Promotion," "Display Only," "Feature Only," and "Feature & Display." Given assumption (3.2), price is calculated as dollar sales for promotion $m$ divided by unit sales for that promotion.

Scanner data reports information regarding product and promotional distribution through a variable known as "All Commodity Volume," or ACV. $ACV_{jt}$ is the percentage of total sales, across all product categories, accounted for by those stores that carry product $j$ in week $t$. This represents the percentage of stores that distribute a particular item. Similarly, $ACV_{jt}^m$ is the fraction of stores where product $j$ has promotional activity $m$. Note that the percentage of stores that carry product $j$ is the sum of its promotional distribution: $ACV_{jt} = \sum_{m \in M} ACV_{jt}^m$.

We use these distribution measures to calculate two variables. First, the model requires $N_{bt}$, the number of brand $b$'s products that are available in time $t$. We also need $\pi_{bt}^m$, the fraction of stores where brand $b$ has promotional activity $m$. Standard scanner data contains sufficient information to construct both $N_{bt}$ and $\pi_{bt}^m$. To calculate the number of products contained within brand $b$'s product line, we add up each product's ACV: $N_{bt} = \sum_{j \in J_t : b_j = b} ACV_{jt}$.

We then add up the fraction of stores where each of brand $b$'s products has promotional activity $m$: $N_{bt}^m = \sum_{j \in J_t : b_j = b} ACV_{jt}^m$. The percentage of stores where brand $b$ has promotion $m$ is calculated as $\pi_{bt}^m = \dfrac{N_{bt}^m}{N_{bt}}$.

While each brand's univariate promotional distribution $\{\pi_{bt}^m\}_{m \in M}$ is calculated in this manner, the joint distribution of each brand's promotions $\{\pi_t^g\}_{g \in G}$ is not reported in aggregate scanner datasets. Additional model restrictions must be imposed in order to estimate the joint promotional distribution, since a continuum of joint distributions is generally possible given a set of univariate distributions.

There is a special case, however, where the joint promotional distribution is uniquely determined by each brand's univariate distribution. Define brand $b$ as having heterogeneous promotions in week $t$ when its promotional activity varies across the stores aggregated in the data. That is, when $\sum_{m \in M} 1_{\pi_{bt}^m > 0} > 1$. The joint promotional distribution is uniquely determined by each brand's univariate distribution if no more than one brand has heterogeneous promotions.

This special case commonly arises for the following reason. In many product categories, including super-premium ice cream, retail chains usually feature or display no more than one brand in the same city and week. All other brands will typically be on "No Promotion" in each of a retailer's stores. At most one brand has heterogeneous promotions in this situation, implying that the joint promotional distribution is uniquely determined by each brand's univariate promotional distribution. This occurs quite often; for 85% of the weekly observations in our dataset, only one joint promotional distribution can occur given each brand's univariate distribution. Our experience indicates this special case frequently arises across a wide range of consumer packaged goods, rather than being specific to super-premium ice cream.

We have two options for dealing with the remaining 15% of observations where two or more brands have heterogeneous promotions. The first is to exclude such observations from the dataset. The model can then be estimated from the remaining data sample where the joint promotional distribution is uniquely determined by the observed marginal distributions. Alternatively, to apply the model to the 15% of observations where the joint distribution is not known, we have to make an additional assumption about functional form to get identification. Specifically, we assume the joint promotional distribution can be constructed from a copula of the marginal distributions. The data is used to estimate the single parameter of the copula jointly with the other model parameters.

Deciding between these two approaches involves the familiar bias-variance tradeoff. The obtained estimates will be more precise if one imposes additional structure that allows the model to be estimated from the full data sample. However, they may be biased if the employed assumptions are invalid. We believe that, on balance, the benefit of exploiting the entire data sample outweighs the cost of imposing the additional model structure detailed below. Of course, those who believe the benefit does not outweigh the cost can instead estimate the model using the subset of the data where the joint promotional distribution is known.

We rely on the following framework. Let each retail chain be composed of a continuum of stores. Brand $b$'s promotional activity in store $s$ during week $t$ is determined by latent variable $\upsilon_{bst}$, which has a standard Normal distribution. This variable is used to assign brand $b$'s promotional activity in store $s$. We assume brand $b$'s promotional activity is weakly increasing in $\upsilon_{bst}$ based on the following rank order of promotional activity, from lowest to highest: "No Promotion," "Display Only," "Feature Only," and "Feature & Display."[7] This allows us to match each value of $\upsilon_{bst}$ to a particular type of promotion using each brand's univariate promotional distribution. Brand $b$'s promotional activity in store $s$ during week $t$ equals $\tilde{m}$ if $\upsilon_{bst} \in [\Phi^{-1}(\sum_{m \in M: m < \tilde{m}} \pi_{bt}^m), \Phi^{-1}(\sum_{m \in M: m \leq \tilde{m}} \pi_{bt}^m)]$, where $\Phi$ denotes the standard Normal cumulative distribution function.

---

[7] It is unclear whether "Display Only" or "Feature Only" is higher ranked. We obtain similar results when we let "Display Only" outrank "Feature Only."

We then use a Gaussian copula to specify the joint distribution of each brand's latent variable $v_{bst}$. Define $v_{st} = \{v_{bst}\}_{b \in B}$, where vector $v_{st}$ is mean-zero and i.i.d. Normally distributed with variance matrix $\Omega$. Denote the probability distribution function of $v_{st}$ by $\phi(v_{st}; \Omega)$. To minimize the number of estimation parameters, we assume $\Omega$ has identical off-diagonal elements $\rho \in [0,1]$ and unit values along the main diagonal.[8]

Parameter $\rho$ represents retailer strategy regarding how products are jointly promoted across stores. Retailers independently set each brand's promotional activity when $\rho$ equals zero. As $\rho$ increases, the promotional activity of competing brands becomes more positively correlated. That is, in stores where a retailer chooses a high level of promotional activity for one brand, for larger values of $\rho$ it more frequently chooses a high level of promotion for the other brands in those stores.

This framework provides sufficient structure to calculate the joint promotional distribution $\{\pi_t^g\}_{g \in G}$. To calculate $\pi_t^g$ for each $g \in G$, we numerically integrate $\phi(v_{st}; \Omega)$ over the range of values where the promotional activity of each brand $b \in B$ equals $m_b(g)$.

$$(3.12) \quad \pi_t^g = \int_{Y_t^g} \phi(v_{st}; \Omega) dv_{st},$$

where $Y_t^g = \{v_{st} : v_{bst} \in [\Phi^{-1}(\sum_{m \in M: m < m_b(g)} \pi_{bt}^m), \Phi^{-1}(\sum_{m \in M: m \leq m_b(g)} \pi_{bt}^m)], \forall b \in B\}$

To summarize, this framework uses each brand's univariate promotional distribution to choose a joint promotional distribution from a family characterized by estimation parameter $\rho$. We estimate $\rho$ jointly with the other demand parameters via maximum likelihood (see the following subsection). As discussed below, $\rho$ is identified by how variation in this parameter

---

[8] Parameter $\rho$ does not vary over time and is identical across retailers. We make this simplifying assumption since only 15% of the dataset's observations identify the joint promotional distribution. A more flexible specification can be employed in situations where it is practical to do so.

impacts predicted market shares. It is not possible to estimate $\rho$ prior to solving for the other demand parameters, since predicted market shares cannot be computed without them.

However, one can use the following two-stage estimation procedure to solve for $\rho$ after using a subset of the data to estimate the other demand parameters. First, restrict the dataset to the 85% of observations where no more than one brand has heterogeneous promotions. Since the joint promotional distribution is uniquely determined by each brand's univariate distribution for these observations, predicted market shares do not depend on $\rho$. All demand parameters except $\rho$ can be estimated using this restricted dataset.

The 15% of observations excluded from the first-stage estimation can then be used to estimate $\rho$. Parameter $\rho$ determines the distribution of promotions across stores, which ultimately affects predicted market shares. This is the case since consumer purchase decisions depend on whether competing brands are promoted in the same or different stores. For example, a feature advertisement may have a smaller impact on a brand's sales when a competing brand is also being featured.

This two-stage procedure is inefficient, since the first-stage estimation relies only on a subset of the available data. That is why we jointly estimate $\rho$ with the other demand parameters. Nonetheless, since it is possible to identify $\rho$ from one subset of the data, and the other demand parameters from the remaining observations, this procedure demonstrates $\rho$ is separately identified.

Estimation

Before estimating the disaggregated promotion model, we must first specify the relationship between unit sales $\hat{q}_{bt}^m$ predicted by the model, and unit sales $q_{bt}^m$ reported in the data. Berry (1994) develops an approach where an aggregate error term $\varepsilon_{bt}^m$ is included in the utility function. Berry relies on an inversion method in which the set $\{\varepsilon_{bt}^m\}_{\substack{b \in B \\ m \in M}}$ is solved for such that, across all brands $b$ and promotions $m$, the unit sales predicted by the model exactly equals observed quantity sold. This specification is theoretically appealing since the error structure is integrated within the utility-based demand model.[9]

This approach has two drawbacks, however. First, the proposed inversion method is computationally intensive. Second, it requires a strong belief that the "correct" model is being employed; Berry discusses how his inversion method is sensitive to model mis-specification. In particular, it performs poorly when there is measurement error in unit sales. While the scanner datasets produced by the major vendors are relatively high quality, they are still subject to measurement problems. In addition to the usual misreporting and data processing errors, data vendors such as ACNielsen and IRI extrapolate aggregate sales from a sample of stores. This leads to measurement error in reported unit sales.

Instead of including the error term as part of the utility function, one might instead model measurement error in log unit sales, $\ln q_{bt}^m = \ln \hat{q}_{bt}^m + \varepsilon_{bt}^m$. This approach is less theoretically appealing since the error term is not linked to the consumer utility function. It also makes the strong assumption that measurement error is the only form of model uncertainty. However, this specification has the practical advantage of being less computationally complex than the Berry inversion method. Although we recognize the theoretical appeal of including the error term in the utility function, we believe the second approach's robustness to measurement error and reduced computational complexity outweighs this benefit. Nevertheless, since other researchers may have different preferences in this regard, it is important to recognize there is nothing intrinsic to the model that prevents $\varepsilon_{bt}^m$ from being directly included in the utility function.

All of the model's parameters are jointly estimated via maximum likelihood. We define $\varepsilon_{bt}^m = \ln q_{bt}^m - \ln \hat{q}_{bt}^m$ and assume $\varepsilon_{bt}^m$ is mean-zero and independently Normally distributed. Inspection of the data revealed that $\varepsilon_{bt}^m$ is heteroskedastic, having a higher variance when a brand is promoted in only a small fraction of stores. We account for this by modeling the log

---

[9] A second advantage is that including the error term within the utility function facilitates use of instrumental variables. This is not an issue here since we lack valid instruments. Supplemental material available upon request details the reasons why valid instruments do not exist for this product category.

variance as a second order polynomial in $\pi_{bt}^m$, the fraction of stores where brand $b$ has promotional activity $m$.

$$(3.13) \quad Var(\varepsilon_{bt}^m) = e^{\alpha_0 + \alpha_1 \pi_{bt}^m + \alpha_2 (\pi_{bt}^m)^2}$$

Denote the probability distribution function of $\varepsilon_{bt}^m$ by $\phi(\varepsilon_{bt}^m, Var(\varepsilon_{bt}^m))$. This formulation leads to log-likelihood function $\ln L = \sum_t \sum_{b \in B} \sum_{m \in M : \pi_{bt}^m > 0} \ln \phi(\varepsilon_{bt}^m, Var(\varepsilon_{bt}^m))$.

We employ a similar error specification for the representative store model. The error term is defined as $\varepsilon_{bt} = \ln q_{bt} - \ln \hat{q}_{bt}$, where $\varepsilon_{bt}$ is mean-zero and independently Normally distributed. Each brand has identical promotional activity across all stores in the representative store model. Therefore, the counterpart to equation (3.13) is to assume $\varepsilon_{bt}$ is homoskedastic with variance $\sigma^2$. This leads to log-likelihood function $\ln L = \sum_t \sum_{b \in B} \ln \phi(\varepsilon_{bt}, \sigma^2)$.

For both models, we employ Newey-West (1987) standard errors using a lag length of four weeks.[10] That is, error terms that are from periods within four weeks of each other can be arbitrarily correlated. Robustness checks indicated the standard errors are not sensitive to the number of time lags allowed.

Empirical Results

To introduce the data, Table 1 presents the fraction of stores and unit sales that are accounted for by each type of promotion. To give equal weight to each city-chain, we calculate these percentages from variable totals for each one, and then take the average. The table also reports the average promoted price for each brand relative to when it is not on promotion.

Promotional activity plays a significant role in this product category, with promotions accounting for 19% to 31% of unit sales depending on the brand. Unit sales are high, relative to

---

[10] We calculate the variance matrix of the parameter estimates using the standard GMM formulas, where the first order conditions from the log-likelihood function are used as moment conditions. Refer to Hamilton (1994) for discussion of this "quasi maximum likelihood" approach of obtaining robust standard errors.

the percentage of stores, for each level of promotional activity other than "No Promotion." This is due to two distinct effects. First, promotions lead to an outward shift in the demand curve for a given brand. Promotional activity is also associated with a price reduction, with approximately 10% lower prices when on "Display Only," and 30% lower when on "Feature Only" or "Feature & Display." These promotional price reductions are a second factor leading to increased sales.

Table 2 presents parameter estimates for the disaggregated promotion model. Price coefficient $\beta^m$ and intercept $\mu^m$ increase (in absolute value) in the level of promotion $m$, with "No Promotion" the lowest promotional activity, "Display Only" and "Feature Only" intermediate promotions, and "Feature & Display" the highest type of promotional activity. The net impact of these two parameter changes is that a promotion unaccompanied by a price reduction leads to only a small, positive increase in consumer utility (and therefore sales). However, since promotions make consumer utility a steeper function of price, a price reduction accompanied by promotional activity has greater impact than the same price reduction and promotion when separately undertaken.

Table 2 also presents parameter estimates for the representative store model. The parameters for "Display Only" and "Feature & Display" are imprecisely estimated. Table 3 reveals why this is the case. Retailers typically employ these types of promotions in only a small fraction of stores in any given week. For example, when Brand A is on "Display Only" in at least one store in a city-chain, on average 2.5% of stores promote Brand A in this manner. Similarly, on average only 8.9% of stores have a "Feature & Display." The representative store model is unable to isolate the impact of "Display Only" or "Feature & Display" using data aggregated with promotions that are more prevalent. In contrast, the disaggregated promotion model estimates these effects quite precisely.

In addition to the demand estimates presented in Table 2, the disaggregated promotion model has an additional parameter $\rho$. Recall that this parameter is used to estimate the joint distribution of each brand's promotional activity. We obtained the corner solution $\rho = 1$. As a robustness check we re-estimated the model assuming $\rho = 0$.[11] We obtained similar results for the other demand parameters. This insensitivity to the value of $\rho$ does not imply, however, that aggregation bias is not a problem. As discussed earlier, the value of $\rho$ affects only those 15% of weeks where at least two brands have heterogeneous promotions. In contrast, aggregation bias affects the results of the representative store model even when only one brand has heterogeneous promotions. In this case, only one joint promotional distribution can arise given each brand's univariate distribution. The representative store model ignores this information, and instead assumes every consumer observes the average promotional activity across stores. It is much more common for retail chains to promote a single brand than two or more brands at the same time. As such, the disaggregated promotion model requires that we estimate the joint distribution of promotions for a subset of those weeks where at least one brand has heterogeneous promotions across stores. This is why the demand estimates produced by our model are similar regardless of whether $\rho = 0$ or $\rho = 1$, even though aggregation bias significantly impacts the results of the representative store framework.

The first set of estimates in Table 4 presents each brand's own-price elasticity for each type of promotional activity. This is followed by the matrix of cross-price elasticity estimates, calculated when each brand is not on promotion. The third set of results reports the impact of each brand's own promotional activity relative to "No Promotion." All three sets of estimates are evaluated at each brand's average price for the given level of promotion, and are calculated assuming the other brands are not on promotion. This implies the promotional effects shown in the third set of results report the combined effect of being on promotion and undergoing the average price reduction for that promotion.

---

[11] We also considered the following alternative framework. We let $\{\pi_t^g\}_{g \in G}$ be a weighted average of two distributions: the distribution that arises when $\rho = 0$ and the distribution when $\rho = 1$. Using this specification, we obtain the same joint distribution as before, where each brand's promotions are positively correlated to the maximum possible extent.

As mentioned earlier, one key difference between the two sets of results is that the estimates for "Display Only" and "Feature & Display" are imprecisely measured in the representative store model. A second, more critical shortcoming is that the promotional effects in the representative store model are implausibly large. For example, while "Display Only" increases Brand A's sales by 53.1% in the disaggregated promotion model, the representative store model predicts an enormous 1914.5% increase. The magnitude of this effect is not a result of imprecise estimates, since the standard error is "only" 315.7%. The representative store model produces similarly implausible estimates for other brands and types of promotions. This finding is consistent with previous research that concludes data aggregation across stores with heterogeneous promotional activity often leads to overestimation of own-brand promotional effects (Link 1995, Christen et al. 1997).

## IV.   Monte Carlo Analysis

The previous section demonstrates that the disaggregated promotion model generates reasonable demand estimates, while the representative store model does not. Nonetheless, it is impossible to state that the former model is superior without knowing the true parameter values. Therefore, this section uses Monte Carlo analysis to study differences between the two models, specifically whether the poor performance of the representative store model results from inadequate control of promotional heterogeneity across stores. We simulate data using the control variables from the super-premium ice cream data in conjunction with the parameter estimates for the disaggregated promotion model. The constructed data is then used to estimate the disaggregated promotion and representative store models. Since the representative store model generates imprecise results, we must employ a large number of Monte Carlo simulations to calculate accurately the average difference between the true and estimated values. The high computational burden of estimating the random coefficients logit model makes doing so impractical.

We therefore conduct this analysis under three simplifying assumptions that allow the two models to be quickly estimated. These modifications are i) we employ a standard logit specification rather than the random coefficients logit model; ii) error term $\varepsilon_{bt}^m$ is i.i.d. $N(0,\sigma^2)$, rather than heteroskedastic; and iii) $\varepsilon_{bt}^m$ is directly included in the utility function rather than being defined as the difference between actual and predicted log unit sales. Under these simplifications, each model is quickly estimated using the Berry (1994) inversion method. The computational burden is much lower than for the random coefficients models employed earlier since all but one of the model parameters, $\rho$, is estimated by ordinary least squares after the Berry inversion technique has been applied.

We generate the data employed in the Monte Carlo simulations using parameter estimates from this simplified version of the disaggregated promotion model. The disaggregated promotion and representative store models are then estimated using data from 5,000 simulations. Table 5 presents two statistics from the analysis. First, it reports the average percent difference between the estimated elasticities and those calculated at the true parameter values. In addition, we report the standard deviation of the percentage difference across the Monte Carlo simulations. By considering both the mean and the standard deviation, we not only assess whether each model produces accurate results on average, but also whether they deliver precise estimates.

The Monte Carlo results for the disaggregated promotion model are quite accurate. This is not surprising since we generated the data assuming it is the correct model. Of greater interest is the performance of the representative store model. As is the case with our earlier empirical findings, imprecise estimates are obtained for the effects of being on "Display Only" and "Feature & Display." In addition, the average estimate for these effects is often quite far from the true value. By construction, in the Monte Carlo analysis the representative store model suffers from a single mis-specification: it ignores inter-store price and promotional heterogeneity. The similarity of the Monte Carlo results and our empirical findings suggests the poor performance of the representative store model is a result of not controlling for marketing-mix heterogeneity across stores.

The disaggregated promotion model requires we estimate an additional parameter $\rho$, which determines the joint distribution of promotions as a function of each brand's univariate distribution. Table 6 reports the histogram of the $\rho$ estimates from the Monte Carlo simulations discussed above, where the true value of $\rho$ equals one. The table also reports histograms from additional Monte Carlo analyses that assume other values for this parameter (0 and .5). In each case, the median estimate is close to the true value. However, $\rho$ is not estimated that precisely. As discussed earlier, this is because $\rho$ is identified from only the 15% of weekly observations where more than one brand has heterogeneous promotions.

A key assumption of the model formulation developed in section three is that each brand has an identical price across those stores where it has the same promotional activity. This condition is less restrictive than the assumption employed in representative store models, where prices are assumed identical across all stores regardless of their promotional activity. Nonetheless, even this weaker condition might still be violated in empirical applications.

We undertake additional Monte Carlo analysis to investigate whether intra-promotional price heterogeneity leads to aggregation bias. We divide each retail-chain into price zones. While prices vary across stores in different zones, price homogeneity assumption (3.2) holds for stores within each zone. For stores in zone $z$, price for brand $b$ and promotion $m$ in time $t$ is determined by $p_{btz}^m = p_{bt}^m e^{[-\frac{\varphi^2}{2} + \zeta_{btz}]}$, where $\zeta_{btz}$ is i.i.d. $N(0, \varphi^2)$. That is, each brand's expected price is the average conditional on its promotional activity, with the standard deviation of log price equal to $\varphi$. The model is simulated using 100 price zones and $\varphi = 15\%$. This calibration allows for a significant degree of intra-promotional price heterogeneity since the price of super-premium ice cream declines an average of 10% to 30% depending on the type of promotion (see Table 1).

For each Monte Carlo simulation, we add together the sales data from the 100 price zones and then estimate the logit demand models employed above. Table 7 presents the results from this analysis. While the disaggregated promotion model does not perform quite as well as before, a failure to model intra-promotional price heterogeneity leads only to minor bias; across the various estimates, the average percentage difference between the true and estimated values is never greater than 6.8%, and is generally much smaller. The representative store model continues to perform worse. As before, own-price elasticities and promotional effects are imprecisely estimated for "Display Only" and "Feature & Display." In addition, the average impact of these promotions is quite different from the true value. This comparison demonstrates that the disaggregated promotion model is a dramatic improvement over the representative store framework, and can be successfully applied even when there is significant price variation across stores with the same promotional activity.

## V. Conclusion

Demand estimation using aggregate data often leads to biased results. However, only limited solutions for avoiding aggregation bias currently exist. They either have informational requirements that go beyond what is typically available, or fail to fully control for promotional heterogeneity across stores. Due to these shortcomings, practitioners continue to rely on representative store aggregate demand models that ignore inter-store promotional heterogeneity, and which are inconsistent with adding up from consumer-level demand. Previous research demonstrates these are the primary factors leading to aggregation bias.

We show how to avoid these leading determinants of aggregation bias. Our framework generalizes beyond the representative store paradigm by explicitly modeling heterogeneous store types. An aggregate demand model consistent with store-level heterogeneity is constructed by adding up demand across each type of store. This formulation requires the fraction of stores of each type, which we show how to estimate using information included in the scanner datasets produced by the major vendors, ACNielsen and IRI.

The presented empirical application demonstrates how to apply our proposed methodology to extant aggregate demand models. We not only show how to avoid aggregation bias, but also obtain results that are more precisely estimated. This is confirmed by Monte Carlo analysis that demonstrates our framework outperforms a counterpart model based on the representative store paradigm. Our results show there are significant gains to explicitly modeling inter-store marketing-mix heterogeneity when estimating aggregate demand, and using our proposed methodology researchers can easily do so.

**References**

Ackerberg, Daniel A. and Marc Rysman (Forthcoming), "Unobserved Product Differentiation in Discrete Choice Models: Estimating Price Elasticities and Welfare Effects," *RAND Journal of Economics*.

Allenby, Greg and Peter Rossi (1991), "There is No Aggregation Bias: Why Macro Logit Models Work," *Journal of Business and Economic Statistics*, 1-14.

Berry, Steven T. (1994), "Estimating Discrete-Choice Models of Product Differentiation," *RAND Journal of Economics*, 242-262.

Blattberg, Robert, Richard Briesch, and Edward Fox (1995), "How Promotions Work," *Marketing Science*, G 122-132.

Boatwright, Peter, Sanjay Dhar, and Peter E. Rossi (2004), "The Role of Retail Competition, Demographics and Account Retail Strategy as Drivers of Promotional Sensitivity," *Quantitative Marketing and Economics*, 169-190.

Bucklin, Randolph and Sunil Gupta (1999), "Commercial Use of UPC Scanner Data: Industry and Academic Perspectives," *Marketing Science*, 247-273.

Christen, Markus, Sachin Gupta, John Porter, Richard Staelin, and Dick Wittink (1997), "Using Market-Level Data to Understand Promotion Effects in a Nonlinear Model," *Journal of Marketing Research*, 322-334.

Chung, Chanjin and Harry Kaiser (2002), "Advertising Evaluation and Cross-Sectional Data Aggregation," *American Journal of Agricultural Economics*, 800-806.

Cotterill, Ronald W. and Pierre O. Samson (2002), "Estimating a Brand-Level Demand System for American Cheese Products to Evaluate Unilateral and Coordinated Market Power Strategies," *American Journal of Agricultural Economics*, 817-823.

Hamilton, James (1994), *Time Series Analysis*. Princeton, NJ: Princeton University Press.

Hausman, Jerry A. and Gregory K. Leonard (2002), "The Competitive Effects of a New Product Introduction: A Case Study," *Journal of Industrial Economics*, 237-263.

Krishnamurthi, Lakshman, S.P. Raj, and Raja Selvam (1990), "Statistical and Managerial Issues in Cross-Sectional Aggregation," Working Paper, Northwestern University.

Krishnamurthi, Lakshman, Raja Selvam, and Michaela Draganska (2000), "Inference Bias in Cross-Sectional Aggregation," Working Paper.

Lewbel, Arthur (1992), "Aggregation with Log-Linear models," *Review of Economic Studies*, 633-642.

Link, Ross (1995), "Are Aggregate Scanner Data Models Biased?," *Journal of Advertising Research*, RC 8-12.

Nevo, Aviv (2000a), "Mergers with Differentiated Products: The Case of the Ready-to-Eat Cereal Industry," *RAND Journal of Economics*, 395-421.

Nevo, Aviv (2000b), "A Practitioner's Guide to Estimation of Random-Coefficients Logit Models of Demand," *Journal of Economics and Management Strategy*, 513-548.

Newey, Whitney and Kenneth West (1987), "A Simple Positive Semi-Definite, Heteroskedasticity and Autocorrelation Consistent Covariance Matrix," *Econometrica*, 703-708.

Perloff, Jeffrey M. and Michael B. Ward (2003), "Welfare, Market Power, and Price Effects of Product Diversity: Canned Juices," Working Paper.

Theil, Henri (1954), *Linear Aggregation of Economic Relations*. Amsterdam: North-Holland.

Train, Kenneth (1999), "Halton Sequences for Mixed Logit," Working Paper, UC Berkeley.

## Table 1
## Summary Statistics

**Brand A**

|  | No Promotion | Display Only | Feature Only | Feature & Display |
|---|---|---|---|---|
| % of Unit Sales | 81.5% | 0.7% | 15.6% | 2.2% |
| % of Stores | 92.6% | 0.4% | 6.5% | 0.5% |
| Avg. Normalized Price | $1.00 | $0.89 | $0.74 | $0.73 |

**Brand B**

|  | No Promotion | Display Only | Feature Only | Feature & Display |
|---|---|---|---|---|
| % of Unit Sales | 68.8% | 1.1% | 25.7% | 4.4% |
| % of Stores | 87.6% | 0.6% | 10.6% | 1.2% |
| Avg. Normalized Price | $1.00 | $0.90 | $0.71 | $0.70 |

**Brand C**

|  | No Promotion | Display Only | Feature Only | Feature & Display |
|---|---|---|---|---|
| % of Unit Sales | 76.0% | 0.5% | 20.1% | 3.4% |
| % of Stores | 89.2% | 0.3% | 9.6% | 0.9% |
| Avg. Normalized Price | $1.00 | $0.91 | $0.75 | $0.76 |

**Brand D**

|  | No Promotion | Display Only | Feature Only | Feature & Display |
|---|---|---|---|---|
| % of Unit Sales | 77.0% | 0.6% | 20.0% | 2.4% |
| % of Stores | 93.1% | 0.2% | 6.2% | 0.5% |
| Avg. Normalized Price | $1.00 | $0.91 | $0.68 | $0.66 |

*Notes*: N=3,520, corresponding to a panel of 11 city-chain combinations, 80 weeks, and 4 brands. After disaggregating by promotional activity, the dataset contains 4,332 observations.

## Table 2

## Parameter Estimates

### Model 1: Disaggregated Random Coefficients Logit Model

Mean Coefficients:

|  | Intercept | Price |
|---|---|---|
| No Promotion |  | **-0.61** |
|  |  | (0.05) |
| Display Only | **0.60** | **-0.79** |
|  | (0.42) | (0.12) |
| Feature Only | **0.83** | **-1.13** |
|  | (0.28) | (0.09) |
| Feature & Display | **2.40** | **-1.34** |
|  | (0.37) | (0.13) |

Standard Deviation of Random Coefficients:

| Price | **0.13** |
|---|---|
|  | (0.08) |
| Display Only | **1.01** |
|  | (0.53) |
| Feature Only | **1.71** |
|  | (0.18) |
| Feature & Display | **1.22** |
|  | (0.42) |

### Model 2: Standard Random Coefficients Logit Model

Mean Coefficients:

|  | Intercept | Price |
|---|---|---|
| No Promotion |  | **-0.59** |
|  |  | (0.04) |
| Display Only | **-0.60** | **1.22** |
|  | (2.77) | (0.91) |
| Feature Only | **1.07** | **-1.03** |
|  | (0.27) | (0.09) |
| Feature & Display | **6.90** | **-2.44** |
|  | (1.58) | (0.61) |

Standard Deviation of Random Coefficients:

| Price | **0.06** |
|---|---|
|  | (0.10) |
| Display Only | **1.29** |
|  | (0.64) |
| Feature Only | **0.96** |
|  | (0.28) |
| Feature & Display | **0.04** |
|  | (0.87) |

*Notes*: Standard errors are reported in parentheses. Model 1: N=4,332, log-likelihood=-335.87, and RMSE=.34. Model 2: N=3,520, log-likelihood=456.34, and RMSE=.21.

Table 3

**Average Percentage of Stores on Promotion**

|         | No Promotion | Display Only | Feature Only | Feature & Display |
|---------|--------------|--------------|--------------|-------------------|
| Brand A | 93.0%        | 2.5%         | 69.9%        | 8.9%              |
| Brand B | 98.2%        | 4.9%         | 87.0%        | 12.6%             |
| Brand C | 90.2%        | 2.3%         | 78.5%        | 11.1%             |
| Brand D | 99.4%        | 5.6%         | 89.9%        | 12.0%             |

*Notes*: For each calculation, the data sample is restricted to those observations where at least one store has the given type of promotional activity for that brand.

## Table 4

### Estimated Own- and Cross-Brand Effects

#### Disaggregated Random Coefficients Logit Model

**Own-Price Elasticity by Promotion**

| | No Promotion | Display Only | Feature Only | Feature & Display |
|---|---|---|---|---|
| Brand A | -1.62 (0.07) | -1.87 (0.24) | -1.84 (0.15) | -2.27 (0.23) |
| Brand B | -1.66 (0.06) | -1.97 (0.24) | -1.90 (0.15) | -2.29 (0.22) |
| Brand C | -1.56 (0.07) | -1.81 (0.23) | -1.72 (0.14) | -2.22 (0.22) |
| Brand D | -1.81 (0.08) | -2.32 (0.28) | -2.15 (0.18) | -2.69 (0.25) |

**Cross-Price Elasticities**
In response to a price increase by:

| | Brand A | Brand B | Brand C | Brand D |
|---|---|---|---|---|
| Brand A | -1.62 (0.07) | 0.07 (0.01) | 0.13 (0.03) | 0.02 (0.00) |
| Brand B | 0.21 (0.03) | -1.66 (0.06) | 0.16 (0.03) | 0.03 (0.01) |
| Brand C | 0.13 (0.03) | 0.05 (0.01) | -1.56 (0.07) | 0.02 (0.00) |
| Brand D | 0.16 (0.03) | 0.06 (0.01) | 0.14 (0.03) | -1.81 (0.08) |

**Own-Brand Promotional Effects**

| | Display Only | Feature Only | Feature & Display |
|---|---|---|---|
| Brand A | 53.1% (7.7%) | 108.2% (8.6%) | 214.3% (15.6%) |
| Brand B | 67.4% (10.6%) | 183.3% (10.9%) | 343.8% (27.6%) |
| Brand C | 44.9% (7.4%) | 89.9% (5.5%) | 180.2% (12.2%) |
| Brand D | 69.2% (19.5%) | 299.2% (22.0%) | 527.0% (74.1%) |

#### Standard Random Coefficients Logit Model

**Own-Price Elasticity by Promotion**

| | No Promotion | Display Only | Feature Only | Feature & Display |
|---|---|---|---|---|
| Brand A | -1.75 (0.07) | 0.76 (0.62) | -1.88 (0.16) | -3.24 (0.77) |
| Brand B | -1.66 (0.07) | 1.30 (1.03) | -1.76 (0.15) | -3.14 (0.68) |
| Brand C | -1.60 (0.07) | 0.89 (0.69) | -1.72 (0.16) | -3.20 (0.78) |
| Brand D | -1.85 (0.09) | 1.54 (1.18) | -2.05 (0.17) | -4.30 (0.94) |

**Cross-Price Elasticities**
In response to a price increase by:

| | Brand A | Brand B | Brand C | Brand D |
|---|---|---|---|---|
| Brand A | -1.75 (0.07) | 0.04 (0.01) | 0.07 (0.02) | 0.01 (0.00) |
| Brand B | 0.11 (0.03) | -1.66 (0.07) | 0.10 (0.03) | 0.02 (0.01) |
| Brand C | 0.07 (0.02) | 0.03 (0.01) | -1.60 (0.07) | 0.01 (0.00) |
| Brand D | 0.09 (0.02) | 0.04 (0.02) | 0.08 (0.02) | -1.85 (0.09) |

**Own-Brand Promotional Effects**

| | Display Only | Feature Only | Feature & Display |
|---|---|---|---|
| Brand A | 1914.5% (315.7%) | 125.1% (8.5%) | 922.2% (199.1%) |
| Brand B | 2768.0% (681.9%) | 148.9% (10.9%) | 1324.2% (339.9%) |
| Brand C | 1584.8% (283.0%) | 100.6% (5.7%) | 642.1% (137.1%) |
| Brand D | 8114.0% (2206.6%) | 234.9% (25.1%) | 2792.4% (968.0%) |

*Notes*: Standard errors are reported in parentheses.

## Table 5
### Monte Carlo Results

**Average Percent Difference Disaggregated Logit Model** | | | | | **Average Percent Difference Standard Logit Model** | | | |

#### Own-Price Elasticity by Promotion

| | No Promotion | Display Only | Feature Only | Feature & Display | No Promotion | Display Only | Feature Only | Feature & Display |
|---|---|---|---|---|---|---|---|---|
| Brand A | 0.09% | 0.03% | 0.00% | 0.01% | 0.04% | -0.10% | -0.71% | -8.11% |
| | (3.11%) | (4.32%) | (4.79%) | (5.15%) | (3.77%) | (79.07%) | (6.73%) | (38.48%) |
| Brand B | 0.09% | 0.02% | 0.00% | -0.01% | 0.05% | 0.35% | -0.63% | -7.33% |
| | (3.11%) | (4.31%) | (4.73%) | (5.05%) | (3.77%) | (79.34%) | (6.67%) | (37.92%) |
| Brand C | 0.09% | 0.03% | -0.01% | 0.00% | 0.05% | 0.04% | -0.70% | -7.97% |
| | (3.11%) | (4.33%) | (4.79%) | (5.18%) | (3.77%) | (79.23%) | (6.75%) | (38.70%) |
| Brand D | 0.09% | 0.02% | 0.00% | -0.02% | 0.05% | 0.70% | -0.61% | -6.92% |
| | (3.11%) | (4.31%) | (4.74%) | (5.07%) | (3.77%) | (79.69%) | (6.69%) | (38.10%) |

#### Cross-Price Elasticities
In response to a price increase by:

| | Brand A | Brand B | Brand C | Brand D | Brand A | Brand B | Brand C | Brand D |
|---|---|---|---|---|---|---|---|---|
| Brand A | 0.09% | 0.06% | 0.09% | 0.03% | 0.04% | -0.13% | 0.05% | -0.11% |
| | (3.11%) | (3.42%) | (3.37%) | (3.37%) | (3.77%) | (4.07%) | (4.07%) | (4.00%) |
| Brand B | 0.06% | 0.09% | 0.09% | 0.03% | 0.19% | 0.05% | 0.05% | -0.11% |
| | (3.31%) | (3.11%) | (3.37%) | (3.37%) | (4.01%) | (3.77%) | (4.07%) | (4.00%) |
| Brand C | 0.06% | 0.06% | 0.09% | 0.03% | 0.19% | -0.13% | 0.05% | -0.11% |
| | (3.31%) | (3.42%) | (3.11%) | (3.37%) | (4.01%) | (4.07%) | (3.77%) | (4.00%) |
| Brand D | 0.06% | 0.06% | 0.09% | 0.09% | 0.19% | -0.13% | 0.05% | 0.05% |
| | (3.31%) | (3.42%) | (3.37%) | (3.11%) | (4.01%) | (4.07%) | (4.07%) | (3.77%) |

#### Own-Brand Promotional Effects

| | Display Only | Feature Only | Feature & Display | Display Only | Feature Only | Feature & Display |
|---|---|---|---|---|---|---|
| Brand A | -0.81% | 0.11% | -0.25% | 56.52% | 2.76% | 26.60% |
| | (6.58%) | (3.65%) | (3.41%) | (112.37%) | (5.06%) | (30.58%) |
| Brand B | -0.83% | 0.09% | -0.26% | 60.44% | 2.38% | 23.56% |
| | (6.83%) | (3.55%) | (3.45%) | (121.24%) | (4.99%) | (27.34%) |
| Brand C | -0.92% | 0.11% | -0.26% | 63.77% | 2.91% | 29.06% |
| | (7.51%) | (3.85%) | (3.73%) | (127.13%) | (5.32%) | (34.54%) |
| Brand D | -0.92% | 0.10% | -0.24% | 67.45% | 2.32% | 23.15% |
| | (7.64%) | (3.36%) | (3.39%) | (133.36%) | (4.74%) | (27.27%) |

*Notes*: The table reports the average percent difference between the true and estimated values across the 5,000 Monte Carlo simulations. The standard deviation of the percent difference is reported in parentheses.

Table 6

**Monte Carlo Results, Histograms for Parameter $\rho$**

True Value: $\rho = 0$

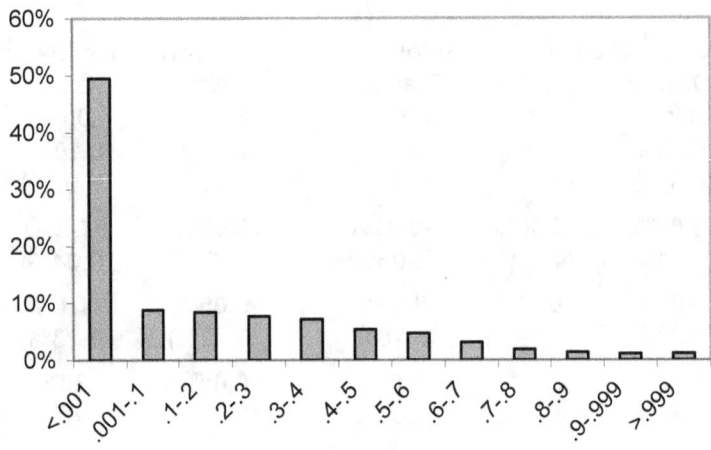

True Value: $\rho = .5$

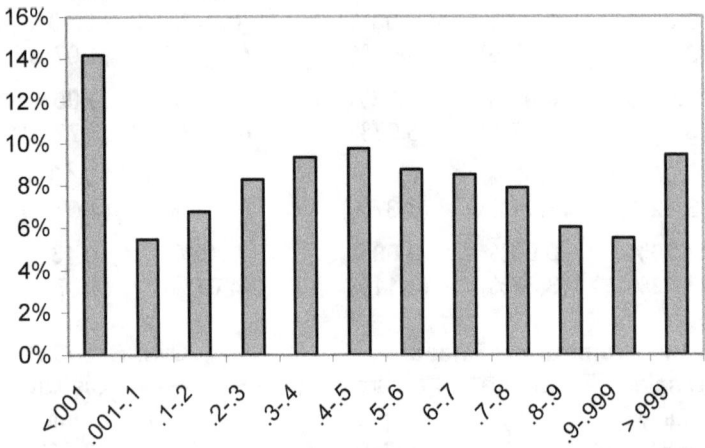

True Value: $\rho = 1$

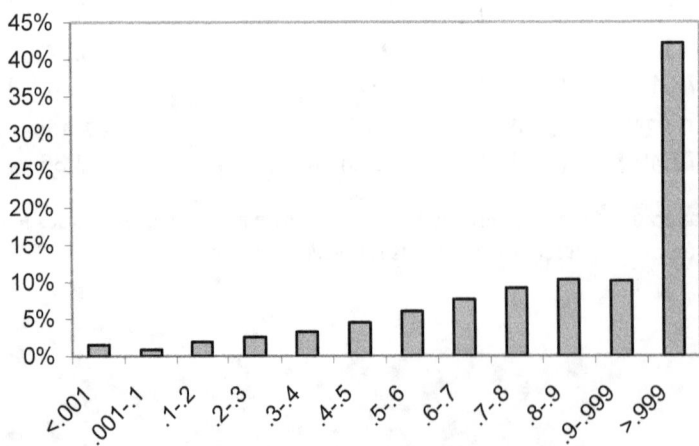

*Notes*: For each value of $\rho$, the table reports histograms from 5,000 Monte Carlo simulations.

## Table 7

## Monte Carlo Results, Allowing for Intra-Promotional Price Heterogeneity

| | Average Percent Difference Disaggregated Logit Model | | | | Average Percent Difference Standard Logit Model | | | |
|---|---|---|---|---|---|---|---|---|
| | **Own-Price Elasticity by Promotion** | | | | **Own-Price Elasticity by Promotion** | | | |
| | No Promotion | Display Only | Feature Only | Feature & Display | No Promotion | Display Only | Feature Only | Feature & Display |
| Brand A | 2.88% (3.32%) | 4.27% (4.63%) | 5.09% (5.29%) | 5.74% (5.80%) | 2.63% (3.97%) | 7.77% (84.30%) | 4.22% (7.58%) | -2.98% (42.34%) |
| Brand B | 2.79% (3.32%) | 4.10% (4.61%) | 4.71% (5.22%) | 5.07% (5.64%) | 2.55% (3.97%) | 8.30% (84.60%) | 3.94% (7.49%) | -2.70% (41.46%) |
| Brand C | 2.88% (3.32%) | 4.27% (4.64%) | 5.09% (5.31%) | 5.72% (5.83%) | 2.63% (3.97%) | 7.94% (84.52%) | 4.24% (7.60%) | -2.84% (42.60%) |
| Brand D | 2.77% (3.32%) | 4.06% (4.61%) | 4.61% (5.23%) | 4.86% (5.64%) | 2.53% (3.97%) | 8.74% (85.04%) | 3.86% (7.50%) | -2.39% (41.62%) |
| | **Cross-Price Elasticities** | | | | **Cross-Price Elasticities** | | | |
| | In response to a price increase by: | | | | In response to a price increase by: | | | |
| | Brand A | Brand B | Brand C | Brand D | Brand A | Brand B | Brand C | Brand D |
| Brand A | 2.88% (3.32%) | -0.18% (3.53%) | -0.43% (3.45%) | -0.48% (3.51%) | 2.63% (3.97%) | -0.60% (4.17%) | -0.68% (4.13%) | -0.83% (4.12%) |
| Brand B | -0.55% (3.45%) | 2.79% (3.32%) | -0.43% (3.45%) | -0.48% (3.51%) | -0.62% (4.11%) | 2.55% (3.97%) | -0.68% (4.13%) | -0.83% (4.12%) |
| Brand C | -0.55% (3.45%) | -0.18% (3.53%) | 2.88% (3.32%) | -0.48% (3.51%) | -0.62% (4.11%) | -0.60% (4.17%) | 2.63% (3.97%) | -0.83% (4.12%) |
| Brand D | -0.55% (3.45%) | -0.18% (3.53%) | -0.43% (3.45%) | 2.77% (3.32%) | -0.62% (4.11%) | -0.60% (4.17%) | -0.68% (4.13%) | 2.53% (3.97%) |
| | **Own-Brand Promotional Effects** | | | | **Own-Brand Promotional Effects** | | | |
| | Display Only | Feature Only | Feature & Display | | Display Only | Feature Only | Feature & Display | |
| Brand A | -3.22% (6.25%) | -5.23% (3.38%) | -6.09% (3.22%) | | 64.74% (110.31%) | -2.65% (4.68%) | 19.53% (28.39%) | |
| Brand B | -3.07% (6.53%) | -4.05% (3.36%) | -5.13% (3.26%) | | 71.06% (121.74%) | -1.77% (4.74%) | 18.28% (26.09%) | |
| Brand C | -4.18% (7.01%) | -5.65% (3.53%) | -6.78% (3.47%) | | 70.77% (122.26%) | -2.95% (4.88%) | 20.73% (31.49%) | |
| Brand D | -4.30% (7.03%) | -3.73% (3.21%) | -4.73% (3.25%) | | 73.70% (126.93%) | -1.53% (4.54%) | 18.45% (26.33%) | |

*Notes*: The table reports the average percent difference between the true and estimated values across the 5,000 Monte Carlo simulations. The standard deviation of the percent difference is reported in parentheses.

www.ingramcontent.com/pod-product-compliance
Lightning Source LLC
Chambersburg PA
CBHW081808170526
45167CB00008B/3377